Tsunamis

Christy Steele

Nature on the Rampage

www.raintreepublishers.co.uk
Visit our website to find out more information about **Raintree** books.

To order:
Phone 44 (0) 1865 888112
Send a fax to 44 (0) 1865 314091
Visit the Raintree Bookshop at www.raintreepublishers.co.uk to browse our catalogue and order online.

First published in Great Britain by Raintree Publishers, Halley Court, Jordan Hill, Oxford, OX2 8EJ, part of Harcourt Education.
Raintree is a registered trademark of Harcourt Education Ltd.

Consultants: Dr George Curtis, Division of Natural Sciences, University of Hawaii at Hilo; David Larwa, Educational Training Services, Brighton, Michigan; Maria Kent Rowell, Sebastopol.

Raintree Editorial: Isabel Thomas and Kate Buckingham
Cover Design: Jo Sapwell (www.tipani.co.uk)

Originated by Dot Gradations
Printed and bound in China by South China Printing Company

ISBN 1 844 21214 9
07 06 05 04 03
10 9 8 7 6 5 4 3 2 1

British Library Cataloguing in Publication Data
Steele, Christy
Tsunamis. - (Nature on the Rampage)
1.Tsunamis - Juvenile literature
I.Title
551.4'7024
A full catalogue for this book is available from the British Library

Acknowledgements
The publishers would like to thank the following for permission to reproduce photographs: NOAA, pp. **1, 4, 8, 20, 23, 24, 26, 29**; Photo Network, p. **10.**

Cover photograph by Oxford Scientific Films/Mary Plage

Contents

What is a tsunami? 5

All about waves 11

Tsunamis in history 19

Science and tsunamis 27

Glossary 30

Addresses and Internet sites 31

Index 32

A tsunami has smashed these buildings into pieces.

What is a tsunami?

A tsunami (pronounced soo-NAM-ee) is a group of fast moving and very deep waves. A tsunami's waves travel out in circles from the point where it starts, like the ripples when a rock is thrown into a pool. The waves move along in a wave train, one wave following another.

A tsunami's waves are usually less than 1 metre high in the open ocean. The waves become much steeper as they reach the coast. Near land, the waves can be as high as 30 metres.

Powerful tsunamis create powerful, killer waves. They wash away anything in their path. They can snap trees like toothpicks, crush houses and toss large boats and rocks on to shore. A tsunami can flood large areas of low-lying land. In the last ten years, more than 4000 people have been killed by tsunamis.

About tsunamis

Tsunamis are huge columns of water that may stretch thousands of metres down into the ocean. They stretch all the way down to the ocean floor.

Tsunamis flow outwards at great speeds. In the deep ocean, they move at up to 800 kilometres (500 miles) per hour. There may be over a hundred kilometres between each tsunami wave. The waves race across thousands of kilometres of open ocean until they reach land. They can travel from one side of an ocean to the opposite shore in less than a day.

Near the coast, where the ocean is not as deep, tsunamis begin to change. Waves bounce off the ocean floor. This makes the tsunami slow down as it nears the coast. The waves of a tsunami are pushed closer together, creating much larger waves. The waves grow taller and taller until they finally smash into the land. A wave 30 cm high in deep water might be 30 metres high when it reaches land.

The shape and height of tsunamis depend on the shape of the ocean floor and coastline where they strike. In some places, offshore

wave slows to 130–160 km/h (80–100 mph) near shore; height of wave can reach 24 metres

calm water level

wave at sea travels at up to 800 km/h (500 mph)

wave motion circles

as water gets shallower, bottom of wave slows down and top of wave begins to topple over

This illustration shows how tsunamis behave as they approach land.

reefs may slightly break the force of a tsunami. A reef is a strip of sand, rocks or coral that rises from the ocean floor almost to the surface of the water. In other places, the sea floor may contain a deep **valley**. A valley is a deep crack in the sea floor. Since a valley is deeper than other areas of the sea floor, the tsunami will not grow as tall there.

A tsunami in Alaska has washed this boat on to the shore.

What a tsunami does

The word tsunami is Japanese for 'habour wave', because of the devastating effects these waves have had on low-lying Japanese coastal towns. Each tsunami acts differently when it strikes. Some tsunamis draw water away from the shore just before they hit. People may see fish flopping on the sand. Other tsunamis wash over land with no warning.

Some people think that tsunamis are over as soon as one wave strikes. However, one tsunami contains many waves. Often the waves that come after the first hit may be even taller and more deadly. The first waves hit the shore and flow back to sea. They then combine with other waves coming to shore. This can make an even larger wave that washes over the land. It may be several hours before all the waves of one tsunami have hit the coast. Sometimes the waves of a tsunami may travel back and forth across the ocean for several days.

Tsunamis can cause a great deal of damage to ocean coasts. Water slowly moves away rock and soil in a process called **erosion**. Erosion usually takes place over a long period of time. But each tsunami causes a lot of erosion all at once. In one moment, a tsunami can strip a beach of all its sand. Tsunamis toss huge rocks from the ocean on to the land. They wash away soil and plants on coastlines.

Tsunamis can also kill people. Some may get washed away in the waves and drown. Flooding from a tsunami may leave people without electricity or clean water to drink.

A wave like this is made when energy travels through water.

All about waves

To understand tsunamis, it is important to understand how normal waves are formed. A wave is a sign that energy is moving from place to place. As a wave moves through the ocean, it does not push the water towards the shore. Instead, the wave disturbs water particles. A particle is a very small part of something. The energy of the wave causes the water particles to tumble around in circles as it passes through. Most normal waves are close to the ocean's surface. Tsunamis are different from other waves. In a tsunami, these circles reach all the way from the sea floor to the ocean's surface. They stretch for thousands of kilometres across the ocean.

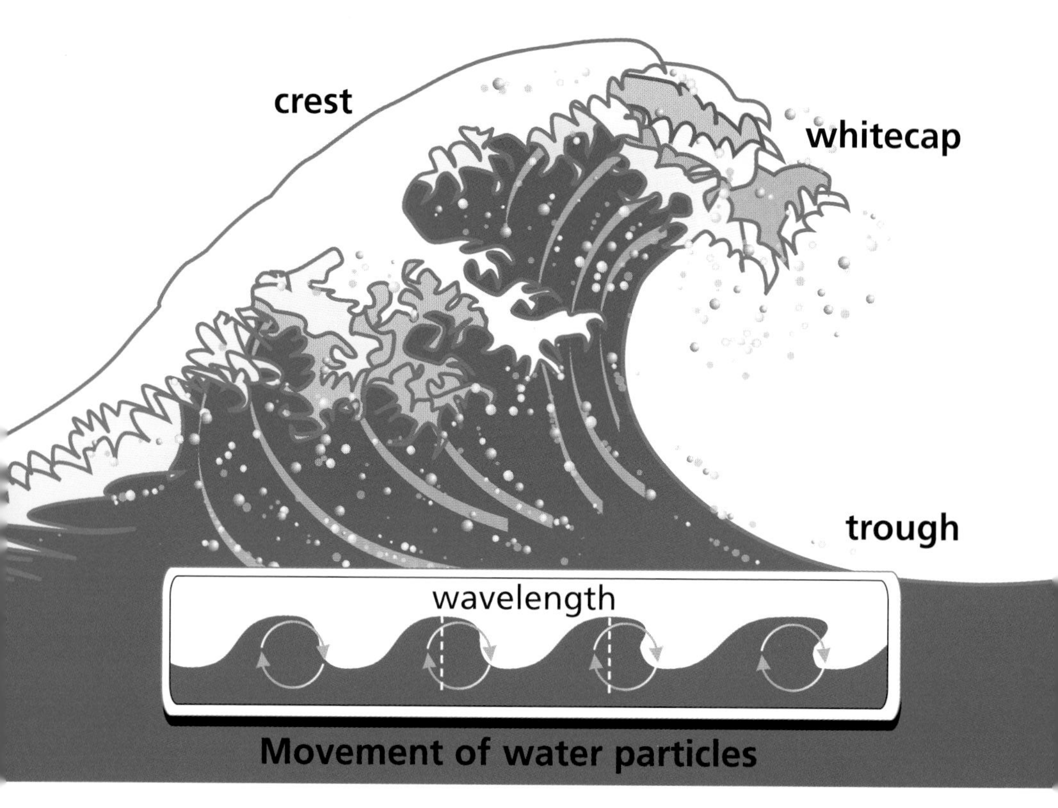

This illustration shows the different parts of a wave.

Parts of a wave

All waves have some parts in common. The **crest** is the highest point of a wave. The **trough** is the lowest point of the wave. The distance from one crest to the next crest is the **wavelength**, or how long the wave is. The time between one crest

Did you know?
Tsunamis were once called tidal waves. Scientists renamed them because the waves have nothing to do with tides.

passing a point and the next crest passing the same point is called the wave period.

Tsunamis have long wavelengths and periods. Their wavelength may be up to 320 kilometres (200 miles) with a period up to 90 minutes. This is enormous compared to the average wind-driven wave, which has a wavelength of only 150 metres (500 feet) and a period of just ten seconds.

When a normal wave gets close to the shore, it slows down from the bottom up. The wave falls over, or breaks, when the front of the wave becomes steeper than the back. Breaking waves often have white foam crests at the top. Tsunamis are unlike other waves. They usually do not break or form white crests. They reach the shore as a wall of water or a wave that arches over before crashing down on the land.

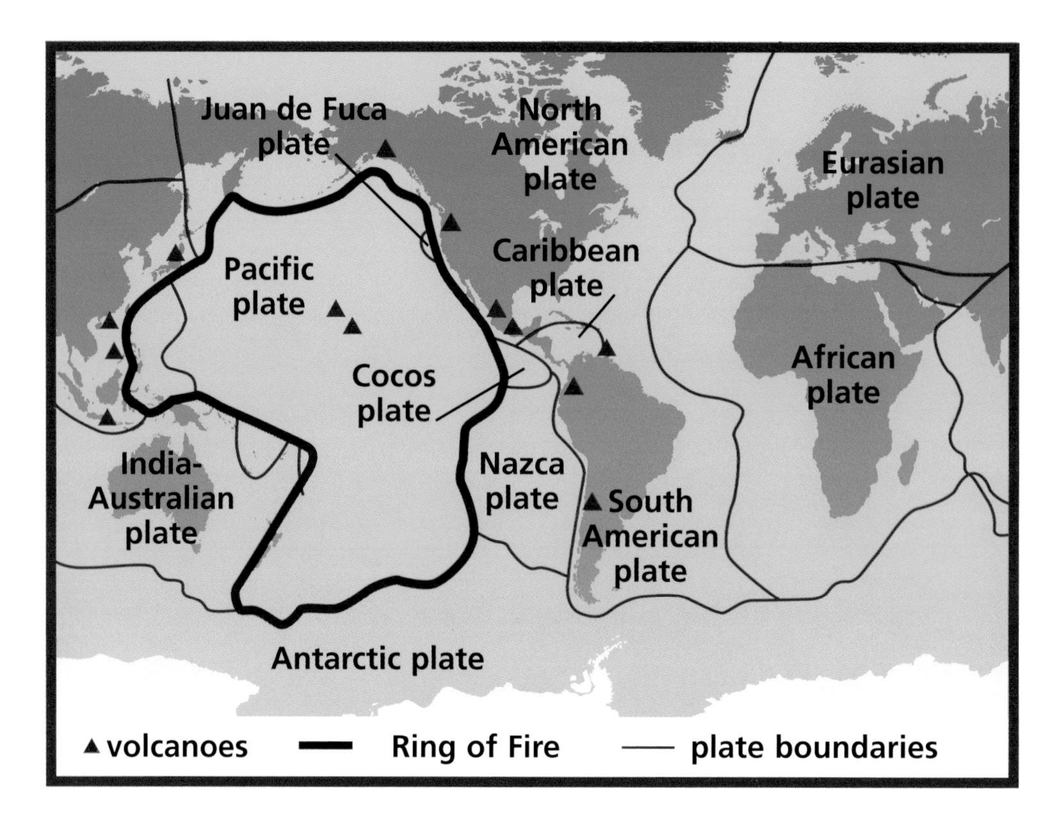

This illustration shows the different plates that make up the Earth's crust.

Causes of tsunamis

Tsunamis start whenever a force pulses energy through a great column of ocean water. The energy moving through the column of water pushes up. It flows outwards from the centre in large circles, sending waves in all directions. The strength of a tsunami depends on several things.

These include the shape of the ocean floor and the depth of the water. It also depends on the amount and movement of energy and the strength of the force that caused it.

Earthquakes are the most common causes of tsunamis. The Earth's crust is divided into giant moving pieces called plates. Some plates carry crust underneath oceans. These plates may crash into each other or slide under each other. The rise or fall of the crust can cause earthquakes under the ocean. This movement disturbs huge columns of water and causes tsunamis. Earthquake tsunamis are often the most dangerous and deadly waves.

Volcano eruptions are another main cause of tsunamis. Hot ash, gas and melted rock called lava flow out of a volcano when it erupts. Volcanoes sometimes cause landslides or huge pieces of volcanic islands to sink back underwater. This causes giant tsunamis. Eruptions can also cause earthquakes that form tsunamis.

Most tsunamis happen around the Pacific Plate. This oceanic plate moves past many other plates. The Pacific Plate's movements cause most of the world's biggest earthquakes and volcanoes to

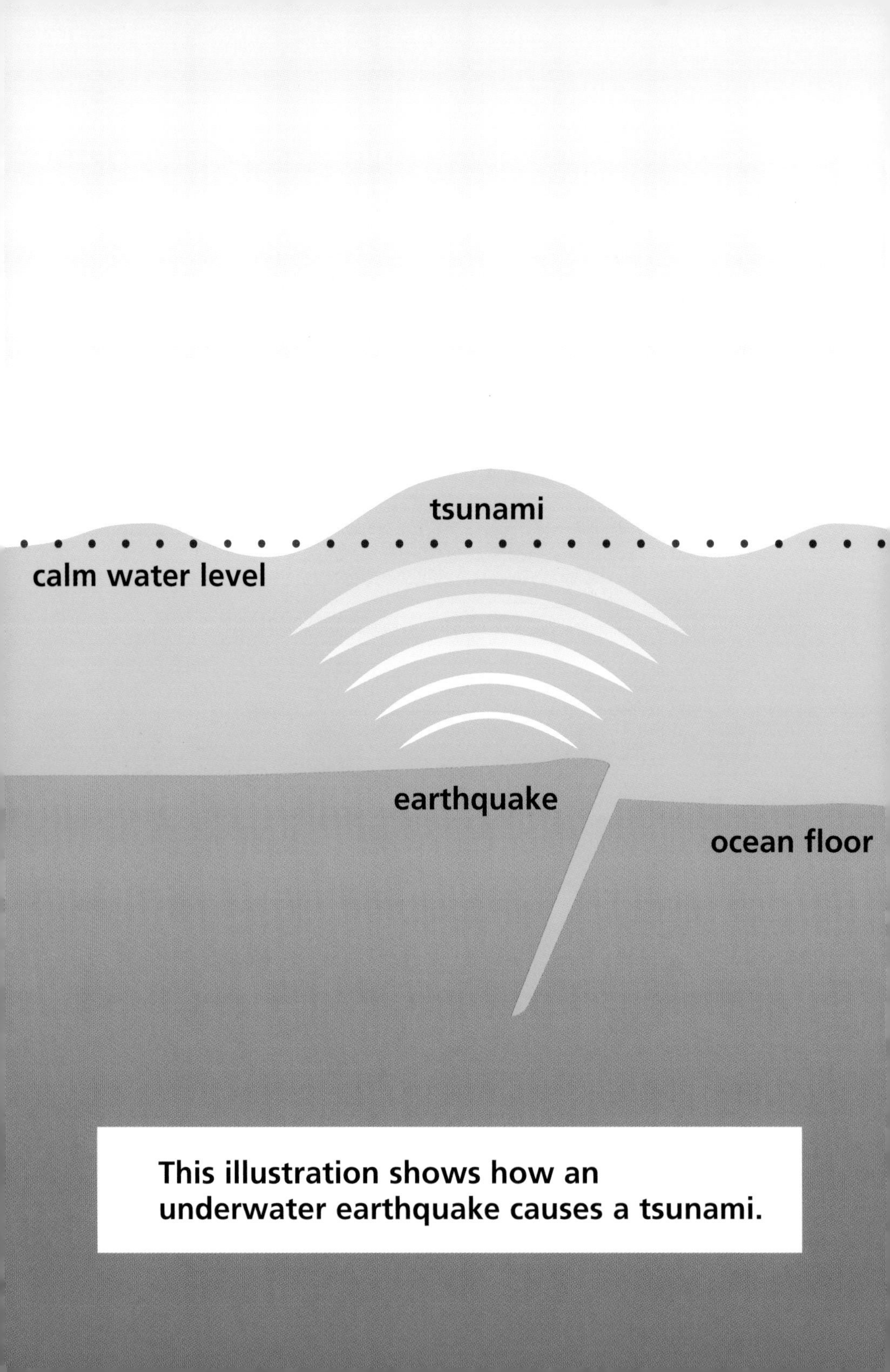

tsunami

calm water level

earthquake

ocean floor

This illustration shows how an underwater earthquake causes a tsunami.

form around its edges. People call the edges of the Pacific Plate the Ring of Fire because of its many volcanoes. Over 25 per cent of the deaths caused by volcanoes over the last 250 years were actually caused by huge tsunamis formed by the eruptions.

Landslides and meteorites

The ocean floor has as many different features as dry land. There are mountains, valleys and flat areas. Sometimes landslides may fall down underwater mountains and cause small tsunamis. An underwater landslide is called a slump.

There are no active volcanoes and very few earthquakes in the UK. But it might be at risk from tsunamis caused by underwater landslides in the North Sea. A slump off the coast of Norway is thought to have caused a huge tsunami that devastated the east coast of Britain 7200 years ago. If this happened today, cities like Edinburgh and Newcastle could be destroyed.

Meteorites falling into the ocean also cause tsunamis. A meteorite is a rock from outer space that has fallen to Earth. This does not happen often – the last major one was 2.16 million years ago.

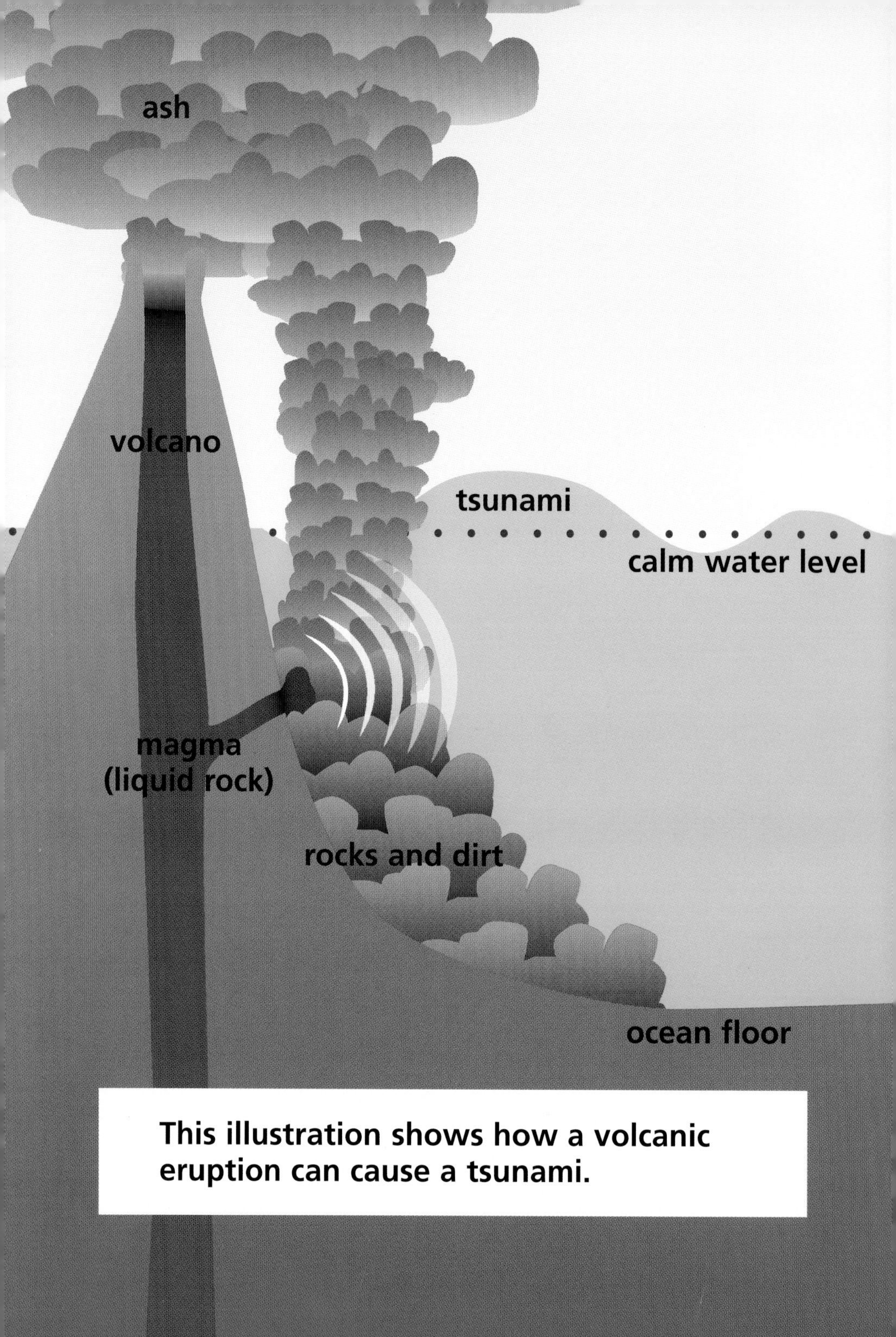

This illustration shows how a volcanic eruption can cause a tsunami.

Tsunamis in history

Tsunamis have happened throughout the Earth's history. People living in the Hawaiian Islands are always in danger from tsunamis. The islands have volcanoes that erupt and cause underwater landslides. The landslides can cause tsunamis.

Hawaiians that lived long ago told this story to explain why tsunamis happen: There was once a magic shark that wanted to sleep forever. He found a quiet place in the ocean and went to sleep. One day, an earthquake woke the shark. He was very angry and threw a wave at the nearest island. Every time a volcano or an earthquake wakes the shark, he gets angry and throws a tsunami towards land.

Now scientists can explain the real reasons that tsunamis happen in different parts of the world.

This is a picture of the 1946 tsunami hitting the shore in Hawaii.

1946 Aleutian tsunami

On 1 April 1946, an earthquake shook the Aleutian Islands of Alaska. The earthquake started a huge tsunami that reached land with waves more than 35 metres high. In minutes, the tsunami hit the shore. It washed boats onto

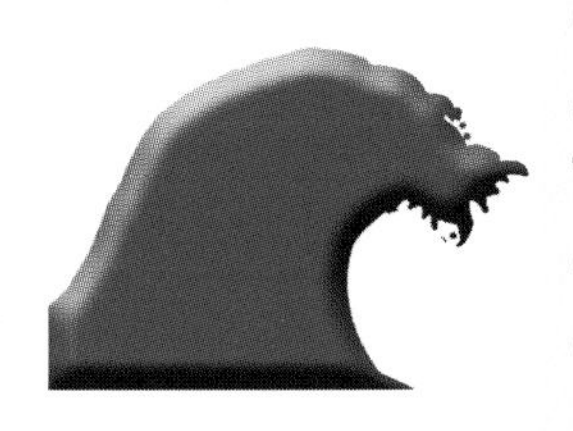

One of the deadliest tsunamis ever recorded struck in 1883. The volcano on the small island of Krakatoa in Indonesia erupted, destroying most of the island. The eruption caused tsunami waves up to 40 metres high that flooded nearby islands, killing more than 36,000 people.

land and smashed houses. It also destroyed a new lighthouse, killing all five lighthouse keepers.

The tsunami raced across the Pacific from Alaska. About five hours later, the tsunami reached Hawaii. People in Hawaii did not know that the tsunami was coming. Some people on the shore heard hissing noises as water ran down away from the shore and was sucked out to sea. Then a huge wave returned and crashed on to the shorelines of the Hawaiian islands. The 12-metre wave covered areas up to 1 kilometre inland and caused US$26 million of damage.

Altogether, more than 165 people died in Alaska and Hawaii. The loss of life and property made the US government start the Pacific Tsunami Warning System. Scientists now work to try and warn people when a tsunami is coming.

1960 Chilean tsunami

On 21 and 22 May 1960, two strong earthquakes shook the ocean floor off the coast of Chile. About fifteen minutes later, the first 10-metre tsunami wave crashed on to the shore. It washed away buildings and sucked their remains out to sea. As wave after wave hit the shore, more buildings and people were destroyed. Up to 2,300 people in Chile died from the earthquakes and the tsunami.

These two earthquakes caused more tsunamis that spread out and struck coastlines all around the Pacific Ocean. Fifteen hours after the earthquakes hit, an 11-metre-high tsunami flooded the city of Hilo, Hawaii, and killed 61 people. Twenty-two hours later, a tsunami hit Japan. Around 200 people died in Japan when the 6-metre waves struck there.

1998 Papua New Guinea tsunami

On 17 July 1998, an underwater earthquake in the south-western Pacific near Papua New Guinea caused a large tsunami. Twenty minutes later, three waves washed over part of the north-eastern coast.

The 1998 Papua New Guinea tsunami flooded this beach, destroying houses.

Waves up to 15 metres high flooded a 40-kilometre (25-mile) stretch of coastline. The tsunami buried people under sand and mud. People drowned or were killed when water smashed their houses. Over 2000 people died from the effects of the tsunami.

A tsunami destroyed this house, which was built in a low-lying area.

Tsunami safety

People can do certain things to stay safe during a tsunami. Scientists from the Pacific Tsunami Warning System call a tsunami watch if an earthquake or landslide takes place in or near the sea. These events might cause a tsunami to strike a certain area. People are told to stay away from the beach during a tsunami watch. They should watch and listen to the television and radio news for more information.

The Tsunami Warning System gives a warning just before a tsunami is about to hit. The warning is broadcast on radio and television. Often, sirens placed along the coast sound warning blasts. The warning broadcasts tell people when a tsunami will reach their area. People living in low-lying places should move to higher ground right away.

Many towns have tsunami **evacuation zones**. Evacuation zones are areas that people must leave if a tsunami is about to strike. People should leave these zones and go to an evacuation site or a safe place during a warning. They should listen to the radio to make sure all the tsunami's waves have hit before they return home.

Scientists study the shape of coastlines like this one where tsunamis often strike.

Science and tsunamis

Today, many scientists study tsunamis. They try to discover where each tsunami started. This helps them work out where waves will travel during future tsunamis so they can warn people.

Scientists also study the shape of coastlines and the ocean floor around them. They put this information into a computer and run a special program. The program shows the scientists how tall tsunamis will grow and what part of the coastline the waves will flood.

Based on the computer program, scientists create tsunami evacuation zones in the places a tsunami is most likely to flood. Some places have laws against putting buildings in tsunami evacuation zones.

Tsunami warning system

In the Pacific, the Tsunami Warning System involves 26 countries. In 1965, these countries agreed to work together to warn each other of tsunamis.

The warning system has stations on islands and land all around the ocean. Each station has instruments called **seismometers**. These measure the strength and location of earthquakes, which may cause tsunamis.

Some stations are buoys, or floats, placed in the ocean itself. The buoys have scientific instruments on them. These instruments can measure the waves of a tsunami as they flow through the deep ocean. Computer chips in the buoys can tell the size of a tsunami by measuring the weight of the water passing over them.

The buoys send this information to a satellite. A satellite is a spacecraft that orbits the Earth. The satellite then sends the information to scientists at the Pacific Tsunami Warning Centre.

This piece of roof is all that remains of a church after it was hit by a tsunami.

New detection equipment helps scientists to give earlier warnings to people living in coastal areas. By doing this, they hope to stop these killer waves from taking people's lives.

Glossary

crest highest point of a wave

earthquake sudden shaking of the ground

erosion moving away of something by water and wind

evacuation zone area people must leave if a tsunami is about to strike

meteorite rock from outer space that has fallen to Earth

reef strip of sand, rocks, or coral that rises from the ocean floor almost to the surface of the water

seismometer instrument that measures the size of an earthquake by measuring its vibrations

trough (TROFF) lowest point of a wave

valley area of low ground between two hills or mountains

volcano vent on the Earth's surface that allows melted rock, ash and gas to erupt out

wavelength distance between one crest of a wave and the next

Addresses and Internet sites

International Tsunami Information Centre
P.O. Box 50027
Honolulu
Hawaii 96850

Pacific Tsunami Museum
P.O. Box 806
Hilo
Hawaii 96721

West Coast and Alaska Tsunami Warning Centre
wcatwc.arh.noaa.gov/subpage1.htm

International Tsunami Information Centre
www.prh.noaa.gov/itic/

Savage Earth – Tsunamis
www.thirteen.org/savageearth/tsunami/

Pacific Tsunami Museum
www.tsunami.org/

Index

Alaska 20, 21

buoy 28, 29

crest 12, 13

earthquake 15, 19, 20, 22, 24, 28
erosion 9
evacuation zone 25, 27

Hawaii 19, 21, 22

landslide 17, 19, 24

meteorite 17

reef 7
Ring of Fire 15

satellite 28
seismometer 28

wave period 13
wavelength 12, 13

trough 12
Tsunami Warning System 21, 24, 25, 28